YOUR KNOWLEDGE HAS VALUE

Akshay Kumar

Harmonic Compensation of Voltage and Current Using UPQC

GRIN Verlag

Bibliografische Information der Deutschen Nationalbibliothek:

Die Deutsche Bibliothek verzeichnet diese Publikation in der Deutschen National-bibliografie; detaillierte bibliografische Daten sind im Internet über http://dnb.d-nb.de/ abrufbar.

Dieses Werk sowie alle darin enthaltenen einzelnen Beiträge und Abbildungen sind urheberrechtlich geschützt. Jede Verwertung, die nicht ausdrücklich vom Urheberrechtsschutz zugelassen ist, bedarf der vorherigen Zustimmung des Verla-ges. Das gilt insbesondere für Vervielfältigungen, Bearbeitungen, Übersetzungen, Mikroverfilmungen, Auswertungen durch Datenbanken und für die Einspeicherung und Verarbeitung in elektronische Systeme. Alle Rechte, auch die des auszugsweisen Nachdrucks, der fotomechanischen Wiedergabe (einschließlich Mikrokopie) sowie der Auswertung durch Datenbanken oder ähnliche Einrichtungen, vorbehalten.

Imprint:

Copyright © 2014 GRIN Verlag GmbH
Druck und Bindung: Books on Demand GmbH, Norderstedt Germany
ISBN: 978-3-656-71744-7

This book at GRIN:

http://www.grin.com/en/e-book/278319/harmonic-compensation-of-voltage-and-current-using-upqc

GRIN - Your knowledge has value

Der GRIN Verlag publiziert seit 1998 wissenschaftliche Arbeiten von Studenten, Hochschullehrern und anderen Akademikern als eBook und gedrucktes Buch. Die Verlagswebsite www.grin.com ist die ideale Plattform zur Veröffentlichung von Hausarbeiten, Abschlussarbeiten, wissenschaftlichen Aufsätzen, Dissertationen und Fachbüchern.

Visit us on the internet:

http://www.grin.com/

http://www.facebook.com/grincom

http://www.twitter.com/grin_com

Harmonic Compensation of Voltage and Current Using UPQC

Akshay kumar

Abstract- **In this paper a novel control method based on Synchronous Reference Frame Theory (SRFT) is proposed to compensate power quality problems through a three- phase Unified Power Quality Conditioner (UPQC) under unbalanced and distorted load conditions. The performance of the proposed system has been verified using MATLAB-SIMULINK and are discussed in detail in this paper.**

Keywords- Phase Locked loop (PLL), Power Quality (PQ), Synchronous reference frame (SRF), Unified Power Quality Conditioner (UPQC), voltage and current harmonics mitigation.

I. INTRODUCTION

With the recent trend of using non-linear and power electronics loads and electronically switched devices in distribution system results in power quality (PQ) problems, such as voltage sags, voltage swells, flickers, interruption, and imbalance of current have become a serious problem. This has attracted the concentrations of many researchers towards the solution of this problem. The custom power devices based on Voltage source converter (VSC) are becoming popular to mitigate these power quality problems. A series converter (also known as Dynamic Voltage Restorer) is able to mitigate the harmonics and distortion in supply voltage such as voltage sag/swell resulting in a regulated voltage across the sensitive/ critical electronics load. A shunt converter (also known as DSTATCOM) is used to compensate for distortion and imbalance in a load resulting a balanced sinusoidal current can flow through the feeder. One new and very promising solution to power quality problem is Unified Power Quality Conditioner (UPQC). The UPQC is a custom power device and it is a combination of both series and shunt converter connected back to back through a common dc link. UPQC can compensate both voltage and current harmonics such as voltage sag/swell and current disturbances. It can also control the power flow and can improve voltage profile [1-].

In this paper a new control technique for the SRF based UPQC is presented, without measuring transformer voltage and load current resulting an improvement of system performance. The proposed control technique has been evaluated and tested under unbalanced and distorted load conditions using MATLAB/SIMULINK software.

II. UPQC

The three phase UPQC consists of two voltage source inverters (VSI) connected back to back to a common DC link. One of these two VSI is connected in series with the ac line while the other VSI is connected in shunt with the same line. The main purpose of the series active filter is harmonic isolation between a sub transmission system and distribution system. In addition series active filter has the capability to mitigate voltage imbalance, stroke flicker compensation as well as voltage regulation and harmonic compensation at PCC. The block diagram of UPQC is shown in Fig. 1.

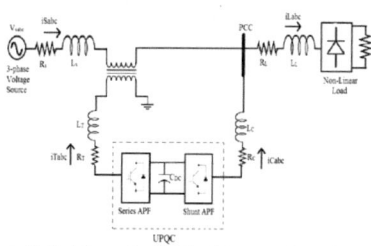

Fig. 1. Unified Power Quality Conditioner configuration

The shunt active filter is used to absorb current harmonic, compensates reactive power and negative sequence current as well as regulates the dc link voltage between two active filters. Two passive filters remove switching frequency harmonics from the output voltage of series converter and output current from shunt inverter.

III. SRF TECHNIQUE

The synchronous reference frame (SRF) method can be used to extract the harmonics contained in supply voltages or currents. For the current harmonic compensation the three phase distorted currents are first converted into two phase stationary coordinated using α-β transformation same as the p-q theory. After that, the stationary frame quantities are transferred into synchronous rotating frame using cosine and sine functions obtained from PLL. The PLL provides the synchronization with supply voltage and current. The harmonics and fundamental components are separated easily by passing the signal from a Low Pass Filter (LPF).

1

After that, the fundamental components transferred back to a-b-c frame using inverse park transformation. The a-b-c to d-q-o transformation is known as park transformation.

IV. UPQC CONTROL ALGORITHM

The proposed UPQC control block diagram in shown in fig.2, sensing three-phase source voltage and current and load voltages along with the dc- link voltages are adequate to compute the reference switching signals in the UPQC.

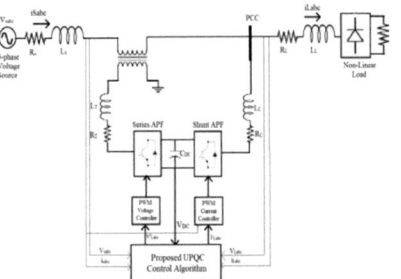

Fig.2. Proposed UPQC control block diagrams

A. Reference Voltage Signal Generation for Series APF

The function of the series APF is to compensate the voltage distribution in the source side, which is due to the fault in the distribution line at PCC [6]. The series APF control algorithm calculates the reference value to be injected by the series APF transformers, comparing the direct axis (d- axis or positive sequence component) with the load side voltage, as shown in fig. no. 3.

The equation (1) shows the transformation of supply voltage V_{Sabc} into d-q-0 coordinates [3-4].

$$
\begin{bmatrix} V_{So} \\ V_{Sd} \\ V_{Sq} \end{bmatrix} =
$$

$$
\sqrt{\frac{2}{3}} \begin{bmatrix} \frac{1}{\sqrt{2}} & \frac{1}{\sqrt{2}} & \frac{1}{\sqrt{2}} \\ \sin(\omega t) & \sin(\omega t - \frac{2\pi}{3}) & \sin(\omega t + \frac{2\pi}{3}) \\ \cos(\omega t) & \cos(\omega t - \frac{2\pi}{3}) & \cos(\omega t + \frac{2\pi}{3}) \end{bmatrix}
\tag{1}
$$

The voltage of d axis (V_{Sd}) given in equation (2) consist of average and oscillating components of source voltages ($\overline{V_{Sd}}$ and $\widetilde{V_{Sd}}$).

The averaging component of voltage $\overline{V_{Sd}}$ is calculated by using second order Low Pass Filter (LPF).

$$
V_{Sd} = (\overline{V_{Sd}} + \widetilde{V_{Sd}})
\tag{2}
$$

The load side reference voltages V^{*}_{Labc} are calculated as given in equation (3). The switching pulses for IGBT (sinusoidal PWM controller) are assessed by comparing reference voltages (V^{*}_{Labc}) with load voltages (V_{Labc}).

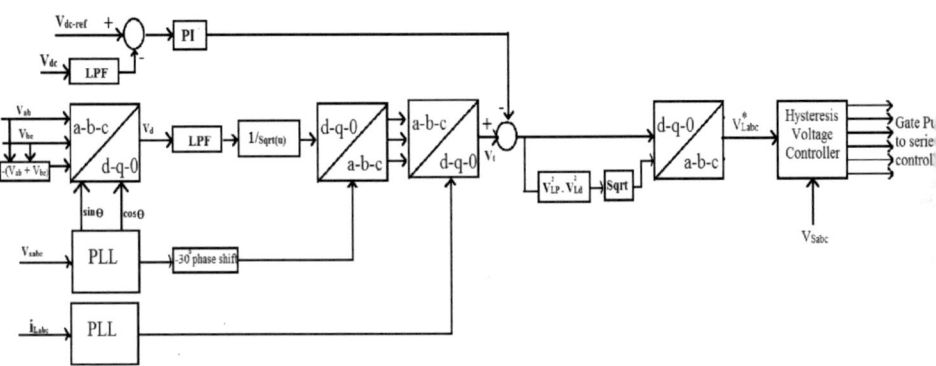

Fig. 3. UPQC Series controller

$$\begin{bmatrix} V_{Sa} \\ V_{Sb} \\ V_{Sc} \end{bmatrix} = \sqrt{\frac{2}{3}} \begin{bmatrix} 1/\sqrt{2} & \sin(\omega t) & \cos(\omega t) \\ 1/\sqrt{2} & \sin(\omega t - 2\pi/3) & \cos(\omega t - 2\pi/3) \\ 1/\sqrt{2} & \sin(\omega t + 2\pi/3) & \cos(\omega t + 2\pi/3) \end{bmatrix} \quad (3)$$

These produced three-phase load reference load voltages are compared with load line voltages and errors are then processed by sinusoidal PWM controller to generate the required switching signals for series APF IGBT switches.

Reference Current Signal Generation for Shunt APF

The shunt APF used to compensate the current harmonics and reactive power generated by non-linear load [7]. The proposed SRF shunt APF reference source current signal current algorithm uses only source voltages source currents and DC-link voltages, as shown in fig no. 4.

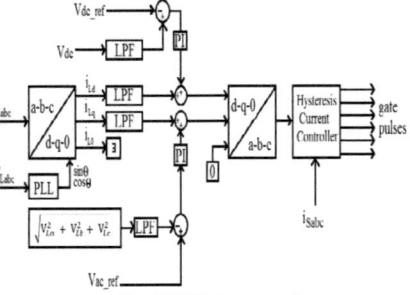

Fig. 4 UPQC shunt controller

Three phase source currents are first transform to d-q-0 coordinates as shown in equation (4).

$$\begin{bmatrix} i_{So} \\ i_{Sd} \\ i_{Sq} \end{bmatrix} = \sqrt{\frac{2}{3}} \begin{bmatrix} 1/\sqrt{2} & 1/\sqrt{2} & 1/\sqrt{2} \\ \sin(\omega t) & \sin(\omega t - 2\pi/3) & \sin(\omega t + 2\pi/3) \\ \cos(\omega t) & \cos(\omega t - 2\pi/3) & \cos(\omega t + 2\pi/3) \end{bmatrix} \quad (4)$$

Instantaneous source currents of d-axis (i_{sd}) given in equation (5) include both oscillating and averaging components of source currents.

$$i_{Sd} = (\overline{i_{Sd}} + \widetilde{i_{Sd}}) \quad (5)$$

The oscillating components consist of harmonics and negative sequence components of source currents. The averaging components consist of positive sequence of currents and correspond to reactive currents. The active power injected by series APF to the power system, compensates the active power loss of the UPQC power circuit, which results in the reduction of dc-link voltage. For this purpose, the dc –link voltage is compared with its reference value (V_{DC}) and the required active current (i_{dloss}) is obtained by PI controller. The fundamental components of reference source current is calculated by adding to the required active current and the source current average component, which is obtain by second order LPF, as given in equation (6).

$$i'_{sd} = i_{dloss} + \overline{i_{Sd}} \quad (6)$$

The source current reference values are calculated as given in equation (7)in order to compensate the harmonics, neutral current unbalance, and reactive power by regulating the dc-link voltage.

$$\begin{bmatrix} i'_{Sa} \\ i'_{Sb} \\ i'_{Sc} \end{bmatrix} = \sqrt{\frac{2}{3}} \begin{bmatrix} 1/\sqrt{2} & \sin(\omega t) & \cos(\omega t) \\ 1/\sqrt{2} & \sin(\omega t - 2\pi/3) & \cos(\omega t - 2\pi/3) \\ 1/\sqrt{2} & \sin(\omega t + 2\pi/3) & \cos(\omega t + 2\pi/3) \end{bmatrix} \quad (7)$$

These reference source current signals are compared with 3-phase source currents and the errors are proceeds by hysteresis band PWM controller to generate the required switching signals for the shunt APF IGBT switches. Simulation model of UPQC is shown in fig. no. 5.

V. SIMULATION RESULTS

This paper describes a new SRF-based control strategy used in the UPQC, which mainly compensates the reactive power along with voltage and current harmonics under non sinusoidal mains voltage and unbalanced load-current conditions. The proposed control strategy uses only loads and mains voltage measurements for the series APF, based on the SRF theory and for shunt measurement load and source current. The simulation results show that, when under unbalanced and nonlinear load-current conditions, the aforementioned control algorithm eliminates the impact of distortion and unbalance of load current on the power line, making the power factor unity. Meanwhile, the series APF isolates the loads and source voltage in sag, unbalanced and distorted load conditions, and the shunt APF compensates reactive power, neutral current, and harmonics and provides three-phase balanced and rated currents for the

3

mains.

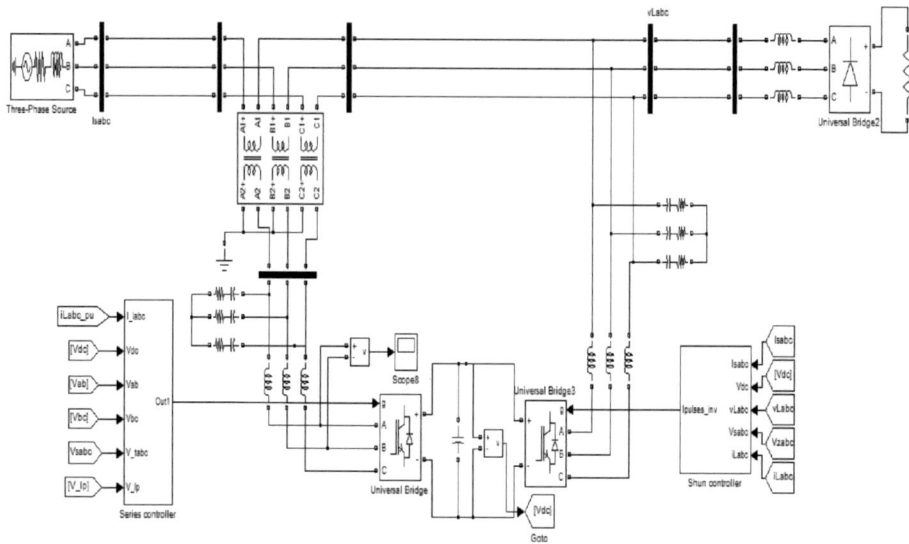

Fi.5. Simulink model of UPQC control diagram

The series APF starts compensating voltage harmonics directly by injecting out of phase harmonic voltage, making load voltage at load distortion free. The voltage injected by series APF is shown in Fig. 6 and during the operation of UPQC DC voltage across the capacitor of back to back VSI is maintained to its reference value. The THD of load current THD is15.17%, while the THD of source current is 2.80%. The load voltage THD is improved form 7.72% to 2.32 %.

VI. CONCLUSION

This paper pointed up a new control strategy for the UPQC system, which mainly compensate voltage sag/swell and current harmonics under distorted mains voltage and load current conditions. The proposed control strategy requires only source current and load voltage measurement for shunt APF based on the indirect current control technique. The synchronous reference frame technique was used by determining mains voltage and filter voltage for series APF so it reduces the number of measurement. The simulation results shown that, the above control algorithms eliminate the impact of distortion of load current on the power line and isolate the loads voltages and source voltage.

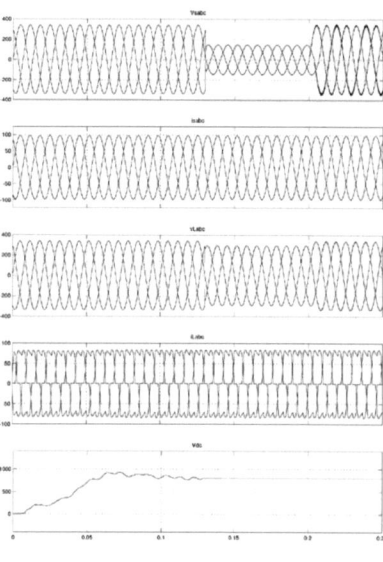

Fig. 6. UPQC wavforms

4

TABLE I
UPQC Experimental and Simulation Parameters

	Parameters		Values
Source	Voltage	Vsabc	415V
	Frequency	f	50Hz
3-phase Non-linear load	Resistive Load	R_{DC}	26 kW
DC link	Voltage	V_{DC}	800V
	DC capacitor	C_{dc}	10000μF
Series APF	AC line inductance	L_{Tabc}	1.5 mH
	Filter Resistor	R_{Tabc}	5 Ω
	Filter Capacitor	C_{Tabc}	2600 μF
	Three phase Transformer	S	3 kVA
Shunt APF	AC Line inductance	L_{Cabc}	0.3 mH
	Filter Resistor	R_{Cabc}	1 Ω
	Filter Capacitor	C_{Cabc}	11.5 mH

REFERENCES

[1] D. Graovac, A. Katic, and A. Rufer, "Power Quality Problems Compensation with Universal Power Quality Conditioning System," IEEE Transaction on Power Delivery, vol. 22, no. 2, 2007W.-K. Chen, Linear Networks and Systems (Book style). Belmont, CA: Wadsworth, 1993, pp. 123–135.

[2] Khadkikar, V.; Chandra, A.; , "A New Control Philosophy for a Unified Power Quality Conditioner (UPQC) to Coordinate Load-Reactive Power Demand Between Shunt and Series Inverters," Power Delivery, IEEE Transactions on , vol.23, no.4, pp.2522-2534, Oct. 2008.

[3] Kesler, M.; Ozdemir, E.; , "Synchronous-Reference-Frame- Based Control Method for UPQC Under Unbalanced and Distorted Load Conditions," Industrial Electronics, IEEE Transactions on , vol.58, no.9, pp.3967-3975, Sept. 2011.

[4] Singh, B.; Solanki, J.; , "A Comparison of Control Algorithms for DSTATCOM," Industrial Electronics, IEEE Transactions on , vol.56, no.7, pp.2738-2745, July 2009.

[5] Singh, B.; Verma, V.; , "Selective Compensation of Power- Quality Problems Through Active Power Filter by Current Decomposition," Power Delivery, IEEE Transactions on , vol.23, no.2, pp.792-799, April 2008.

[6] Lee, G.-M.; Dong-Choon Lee; Jul-Ki Seok; , "Control of series active power filters compensating for source voltage unbalance and current harmonics," Industrial Electronics, IEEE Transactions on , vol.51, no.1, pp. 132- 139, Feb. 2004.

[7] Chandra, A.; Singh, B.; Singh, B.N.; Al-Haddad, K.; , "An improved control algorithm of shunt active filter for voltage regulation, harmonic elimination, power-factor correction, and balancing of nonlinear loads," Power Electronics, IEEE Transactions on , vol.15, no.3, pp.495-507, May 2000.

Akshay Kumar was born in Ghaziabad, India, in 1988. He received the B. Tech (Electrical and Electronics) degree from Ideal Institute of Technology, Ghaziabad (Affiliated to UPTU, Uttar Pradesh), in 2009 and, recently completed his M. Tech (Electrical Power and Energy Systems) from Ajay Kumar Garg Engineering College, Ghaziabad (affiliated to MTU, Uttar Pradesh). He worked in Industry and he has also teaching experience of more than 1 and half year in Engineering College.